The Colorado is tamed. A lake forms back of the dam and the water passes through the cliffs around the huge structure under positive control by gates and valves

Foreword

THE BUREAU OF RECLAMATION, the builder of Hoover* Dam, is a part of the Department of the Interior and one of the principal construction agencies of the Federal government. It has an enviable record as a pioneer in the field of engineering design and a builder of many of the nation's most important structures.

The Bureau has designed and supervised the construction of the world's five largest concrete dams—Hoover on the Colorado River in Arizona and Nevada, Grand Coulee on the Columbia in Washington, Shasta on the Sacramento and Friant on the San Joaquin in California, and Marshall Ford on the Colorado in Texas. On four occasions it has completed the world's highest dams—in 1910 the Shoshone Dam (328′) on the Shoshone River in Wyoming, in 1915 the Arrowrock Dam (349′) on the Boise River in Idaho, in 1932 the Owyhee Dam (417′) on the Owyhee River in Oregon, and in 1935 Hoover Dam (726′) on the Colorado River. ‡

Delegated by law to construct projects for irrigation and related purposes, the Bureau's activities are confined to the arid and semi-arid regions in the seventeen western states. Here it has through careful planning and well executed operations brought water to millions of acres of parched lands, thus providing homes for thousands of settlers, forming the nucleus for thriving communities and increasing the wealth of the nation a half billion dollars annually. On all reclamation projects the cost of the irrigation and power facilities is repaid to the Federal government directly by the water users and through the sale of electrical energy.

Hoover Dam is an outstanding example not only of the ability of the Bureau of Reclamation to construct large and intricate structures but also of the operation of the repayment plan in accord with the Bureau's policy. A discussion of these factors will be found in the following pages.

---

*The name "Boulder Dam" (which is used in this booklet) was changed by Congress in 1947 to Hoover Dam.

‡Many dams have been built since Hoover Dam. The most notable is perhaps Glen Canyon Dam (710′), completed in 1964.

Looking upstream through the deep gorge of Black Canyon, before construction was started

Out in the Black Canyon of the Colorado where the river forms the boundary between Nevada and Arizona, the United States Department of Interior through the Bureau of Reclamation has built Boulder Dam, the highest structure of its kind in the world. Declared by many to be the outstanding engineering feat of the age, the achievement is noteworthy not only by reason of mammoth dimensions, complexity of design and intricateness of construction, but as well by the victory of man over the desert's summer sun, the vagaries of the most treacherous river, and the extreme hazards of great heights on the chasm's sheer walls.

Located in the center of a southwestern desert, the nearest towns to the site are Boulder City, Nevada, eight miles west, and Las Vegas, Nevada, thirty miles to the northwest. U. S. Highway 15 at Las Vegas and 66 at Kingman, Arizona are connected by a highly improved road 113 miles in length which crosses Black Canyon on the dam crest. Main lines of the Union Pacific Railroad pass through Las Vegas and the Santa Fe through Kingman. Boulder City and the dam are connected by bus with both railroads.

COLORADO RIVER

In the irresistible march of the Colorado from the melting snows of the high Rockies to the blue-green waters of the Gulf of California, the river has worn its way through the high plateaus of northern Arizona thus creating magnificent Grand Canyon, cut deep gorges through the Black Mountains at Boulder and Black Canyons approximately 250 miles downstream and in the delta region near the river's mouth has in untold centuries thrown a mountain of silt entirely across the Gulf cutting off the upper end from the sea and forming the basin of Imperial Valley.

Points of interest along the Colorado are located the following distances in miles from the Gulf: Yuma 100, Imperial Dam of the All American Canal 118, Parker Dam at the inlet of the aqueduct to Los Angeles 266, Needles 324, Black Canyon 422, Boulder Canyon 443, Bright Angel crossing in Grand Canyon 695, and the headwaters 1400.

Flood waters cut a new river across the lands of Imperial Valley. The falls swept up the valley at the rate of a mile in three days, destroying forever the farms in its path

Drainage is collected from an area of one-thirteenth of the United States, in Arizona, California, Colorado, Nevada, New Mexico, Utah, and Wyoming. Most of the lands along the banks lie in arid regions where crops can be successfully grown only by irrigation, and the river's waters have become therefore the very life blood of a large part of the southwest. Millions of acres of land have been brought into cultivation along the stream and its tributaries, resulting in the growth of thriving communities, establishment of industries and creation of thousands of homes. However, three-fourths of the river's flow occurs from April to July, when violent floods bring destruction, while in the remaining months of the year the dread spectre of drought is ever present. In 1905 the river broke through the soft banks of the delta region and overflowed into the Imperial Valley, cutting deep gorges in fertile lands, destroying farms and small towns, and threatening inundation of the valley to a depth as great as three hundred feet. Eighteen months of untiring efforts and the expenditure of over two million dollars were required to turn the current back into the old channel. A hundred and fifty miles of levees have been built since that time to hold the river in check, of which less than a third are now effective, the remainder being destroyed by the river's wanderings.

At other times crops have withered and died from lack of sufficient moisture, livestock have been sacrificed and water hauled a hundred miles by truck and train for household use. And always the irrigator was toiling to keep his canals and ditches free from deposited silt or his growing crops from being covered by the semi-liquid muck.

ESTABLISHMENT OF BOULDER CANYON PROJECT

Investigations and surveys by the Bureau of Reclamation had determined at an early date that the problems confronting the land dwellers in the rich valleys of Arizona and California could be solved readily by a large reservoir and high dam at Boulder Canyon or Black Canyon, but the cost was too great to be borne by the lands directly benefited. It was not until the fast growing cities of Southern California supplied a market for domestic water and industrial power, thus providing a means of financing construction, that the project was considered economically feasible.

After several years of arguments between states and legislative battles, the Swing-Johnson Bill became the Boulder Canyon Project Act and was

A part of the town of Mexicali was destroyed by the new river formed in the flood of 1905-1906

signed by President Coolidge on December 21, 1928. The Act provided for the construction of a dam and appurtenant works at Boulder or Black Canyon and the All-American canal to the Imperial Valley at a total cost not to exceed $165,000,000 with the proviso that no appropriations would be made until at least six of the Colorado River Basin states had agreed upon an allocation of the stream flow and contracts had been signed for electrical energy to pay the $126,500,000 cost of construction at the damsite. All state legislatures excepting Arizona, ratified the Santa Fe Compact which distributed the use of the river's waters, the City of Los Angeles, Metropolitan Water District of Southern California, and the Southern California Edison Company signed contracts for electrical energy to pay construction charges within fifty years, and on July 3, 1930, President Hoover signed a bill making the first appropriation available for commencing work.

INVESTIGATIONS AND DESIGNS

Before construction could proceed in earnest, it was necessary to complete designs and drawings, build railroads and highways, erect a transmission line to the canyon, lay out a small city to house the workers and award contracts for labor and materials.

A study of economic factors, diamond drilling in the river channel, and geological and aerial surveys had shown the site at Black Canyon to be more advantageous in practically all particulars. The location was more accessible for railroads and highways, the principal power markets were not so far distant, geological conditions were better, depth to bedrock was less, and a smaller dam would provide the same reservoir capacity.

In order to fulfill the project's four major purposes of flood control, water storage, silt control and generation of electrical energy, the government engineers designed an arch-gravity dam rising 727 feet above bedrock behind which a reservoir would form 585 feet deep, 115 miles in length, with a capacity of 32,350,000 acre-feet—by far the largest artificial lake in the world. The lower 300 to 360 feet of the reservoir depth was allotted as a silt pocket, the next 155 to 215 feet for active storage or regulation and the upper 75 feet for flood control. An impression of the immense volume of water in the full reservoir may be gained from the fact that it is sufficient to cover the state of Connecticut 10 feet deep, or is enough to give every person in the world 5000 gal-

The damsite was diamond drilled from barges

lons. Only about 12,000 acres in the reservoir site was owned by others than the government including the small towns of St. Thomas and Kaolin, Nevada.

To protect the dam and powerhouse from overflow, two immense spillways were placed one on each side of the canyon, upstream from the dam and slightly below the elevation of the dam crest. Each is large enough to carry a flow of 200,000 cubic feet per second, the largest ever recorded in Black Canyon. Outlets for the spillways are down long inclined shafts to river level and thence through 50-foot diameter tunnels (once used for diversion) to the river channel, below the canyon workings.

Withdrawal of water from the reservoir for powerhouse turbines, or downstream irrigation and domestic water requirements, was provided by four graceful intake towers in the reservoir site close to the dam, and systems of huge steel pipes in concrete lined tunnels leading from the towers through canyon walls to the powerhouse, or past that structure to needle valve outlet works. A volume of 125,000 cubic feet per second can be discharged through the intake tower gates and connected lines.

The ultimate powerhouse installation will be fifteen 115,000 horsepower vertical shaft turbines and two of 55,000, or a capacity of 1,835,000 horsepower. Power allocations are 63 per cent to the Southern California power contractors previously mentioned and the remainder in even amounts to the states of Arizona and Nevada. Firm power, available throughout the year, is contracted at 1.63 mills per k. w. h. and secondary power, produced at the convenience of the government, at 0.5 mills.

Detouring the river, while construction was in progress for the dam and powerhouse, was secured by four 50-foot diameter diversion tunnels through the canyon walls, two on each side of the channel, and two cofferdams, one above the other below the dam and powerhouse.

Surveys of canyon walls were extremely hazardous and required unusual methods

A view of Boulder City from the air

PREPARATORY CONSTRUCTION

Moving the immense amounts of materials and heavy equipment required for construction, and the securing of ready access to the canyon activities, necessitated the building of modern railroads and highway connections to trunk lines. Standard gauge tracks were laid by the Union Pacific Railroad, from a point near Las Vegas, to Boulder City; lengthened by the Bureau of Reclamation to the top of the damsite; and later extended by the principal labor contractor to the bottom of Black Canyon and to all its plants, completing the 52 miles of line that served the project. At the peak of construction 300 cars of materials were transported by rail to the dam and appurtenant works every 24 hours.

All important points on the project were linked by paved roads, built to withstand the pounding of 50-ton trucks, 150-passenger transports, and the unending line of project visitors who in one year numbered more than a third of a million and in excess of 40,000 in a single month.

Cheap electrical power for Boulder City and project construction was obtained from a transmission line 222 miles in length and operated at 88,000 volts, which crossed the desert from San Bernardino, California and was erected by the Nevada-California Power Co. under government contract. More than 200,000,000 k. w. h. of electricity were consumed on the project during the construction, costing approximately $2,000,000.00.

Climatic conditions in the desert region are generally ideal for eight months of the year but from May 15 to September 15 shade temperatures in the canyon rise as high as 128° F and the reddish black walls reflect furnace-like waves of heat. A location for a town was therefore chosen on a ridge seven miles from the damsite and 2000 feet above the river, where air temperatures were 10° cooler, and soil conditions for landscaping, to subdue the glare of white sands, were the best obtainable.

Boulder City grew in a year from a desert waste into a community of a thousand homes, a dozen dormitories, four churches, a grade school, shops, stores, restaurants, garages, a 700-seat theater, tourist camps, recreation halls, later a beautiful hotel, and many other trade facilities. Each of the large dormi-

The construction and operation town of Boulder City was built at a location seven miles from Black Canyon

tories of the contractor, Six Companies Inc., was air cooled and heated, and contained 172 single rooms. A mess hall, operated by the Anderson Brothers Supply Co., had seating capacity for 1300, on many days served during 7 times a total of 6000 meals, and more than 6,500,000 meals during the construction period. The contractor's 60-bed hospital was equipped with all modern conveniences and two ambulances and first-aid stations were maintained at the canyon. More than 35,000 cases were treated at the stations and as many as 1,500 in one month.

Water supply for Boulder City, always a problem in desert regions, was first procured from the Colorado River and later from the reservoir. When secured from the river, the water's average silt content of 6,000 parts per million was removed in a settling works near the river's edge, pumped 1800 feet in elevation to a plant in Boulder City where its hardness of 350 p.p.m. average was reduced and impurities removed by filtering and chlorination, before being lifted another 200 feet to the 2,000,000 gallon distribution tank on the hill north of the town.

At the peak of construction, the town's population was 6,000. The workmen came from all states in the union, their

Government residences were built in a permanent manner

average age was 32, and 40 per cent were unmarried. As many as 5250 were directly employed at one time, the monthly payroll exceeding $750,000.00. Based on statistical evidence, it is estimated that upwards of 40,000 persons received their livelihood from the project expenditures through direct labor and the supplying of materials.

All land in and near Boulder City was government owned, the contractors, permittees for business enterprises and non-government residents leasing the ground on which their buildings were located. The area was under strict supervision, and regulations were enforced by Federal Rangers. It is of interest to

The Administration building contains the offices of the Bureau of Reclamation

The 172-room dormitories of the principal contractor were air-cooled

Men were transported to the canyon in motor lorries, some of 150-passenger capacity

note that only one major crime, the robbery of the theater, occurred in the five years of construction, and in this case the criminals were apprehended within two hours of the robbery.

Preparation of drawings and designs and the awarding of contracts to lowest bidders for labor or materials was a continuing process throughout construction. Designs for all features were made in the Denver, Colorado office of the Bureau of Reclamation, and were complete down to the last nut and bolt. Models of structures were made to a scale as large as 1 to 20, research was conducted in many Universities, concrete cylinders as large as 3 feet in diameter were broken in powerful testing machines to ascertain the strength of the cobble-contained concrete. As an example of the completeness of design it is of interest to note that every pound of metal in the structures is connected by copper cable to huge copper grounding mats in the powerhouse tailrace and in the reservoir upstream from the dam. More than 4000 separate drawings are on file in the Boulder City office.

The largest contract, and incidentally the largest awarded by the United States before World War II, was given to Six Companies Inc., of San Francisco on March 11, 1931 for the labor involved in

Men and materials traveled by cableway from canyon rim to locations on canyon walls

the building of the dam, powerhouse and appurtenant works. The contractors' gross earnings amounted to more than $54,000,000. Another large contract was that granted on July 9, 1932 to the Babcock & Wilcox Company of Barberton, Ohio, for furnishing materials, fabricating, and erecting plate steel pipe for the pressure lines leading from the bases of intake towers in the reservoir to the power plant or to outlet works. Gross earnings exceeded $11,500,000. Supplying of materials under other contracts amounted to more than $20,000,000, and freight approximately $8,000,000.

In order to build the dam and accessory features in an efficient and economical manner, and within the time allotted by the government, Six Companies Inc., built highways and railroads, erected machine shops, air compressor plants, garages and warehouses, spanned the canyon first by bridges and later with cableways, constructed a gravel screening plant and two mixing plants, acquired power draglines and shovels, trucks, cars, derricks, cranes, and designed many original appliances for particular needs. Many trucks would carry 16 cubic yards of rock in one load, two were of 50-ton capacity, and others were used to transport 100 or 150 men at one trip. This was undoubtedly the greatest massing of specialized equipment ever witnessed on a construction project.

COMMENCEMENT OF CONSTRUCTION

The general program of construction, all features of which were overlapping, was to drive and line the four diversion tunnels, build the cofferdams, excavate the dam abutments and river channel, build the dam and construct the spillways, intake towers, penstock and outlet systems, outlet works, powerhouse, and switching station. Water for downstream irrigation purposes was first bypassed through the diversion tunnels, then through slide gates in the outer Nevada tunnel when storage was started in the reservoir, and finally through intake towers, outlet works and powerplant turbines.

The first work in the canyon was the building of roads to the damsite

The first work in the canyon was the detouring of the river from its ages-old channel while the dam and powerhouse were built. Four diversion tunnels, two on each side of the river, were driven around the damsite, the average length of each being more than three-quarters of a mile. These were first excavated to a 56-foot diameter and then lined with a 3-foot thickness of concrete. After lining, holes were drilled

Diversion tunnels were driven through the canyon walls around the damsite. The outlet portals of three tunnels are shown

The upper 41 feet of the 56-foot tunnel section was driven in one operation

through the concrete and into the rock, through which a water-cement mixture of grout was injected to fill up all cracks and crevices.

Special apparatus, termed "Jumbos," were designed and built for drilling the tunnel headings, trimming the excavated sections, lining the tunnels, and grouting behind the lining. One of the drilling jumbos carrying thirty drifter-drills and, mounted on a 10-ton truck, would be backed up to a heading, water and air lines connected, and valves opened. The roar seemed deafening as the bits bored their way five, ten, twenty feet into the hard rock. Drilling completed, the holes were loaded with dynamite, equipment removed, wires connected from detonators to firing circuit, and when the switch was thrown the resulting blast shook the canyon walls. Power shovels of 3½-cubic-yard capacity and 10-ton trucks moved up to the face and soon a long line of trucks filed slowly from tunnel portals up winding roads to dump grounds in side canyons.

Each blasting round broke a tunnel

Holes were drilled in the tunnel heading from a "Jumbo" which mounted 30 drills

A 3-foot thickness of concrete was placed in tunnel linings behind massive steel forms

After the tunnels were lined to a 50-foot inside diameter, grout was forced into all cracks and crevices in the surrounding rock

Sand, gravel and cobbles for concrete were secured from a pit ten miles from the damsite

length of 17 feet of 41-foot by 56-foot heading, loosening 2,400 tons of rock. Just a year was required to drive all four tunnels, removing 3,500,000 tons of rock and using approximately one pound of dynamite per ton of excavation.

The first concrete for government construction was poured in the inlet portal of the inner Nevada diversion tunnel on March 5, 1932, and lining of the four bores with a three-foot thickness of concrete was started shortly thereafter. Placing concrete in linings was accomplished by the use of huge steel forms, the one for the sidewall weighing 250 tons for an eighty-foot length. The arch was formed by forcing concrete above the massive forms by compressed air guns operating at 100 pounds pressure per square inch. Lining was practically completed in a year's time, the average placing rate being nearly 1,000 cubic yards a day. Transportation was in two or four-

Materials from the pit were separated into different sizes at the screening and washing plant in Hemenway Wash

cubic-yard buckets by truck from plant to form, and placement was by means of gantry cranes or compressed air guns.

Concrete was produced by the "Lo-Mix" plant of four 4-cubic-yard mixers, situated on a bench 80 feet above the river's edge and 4,000 feet upstream from the damsite. Although dwarfed by the sheer canyon walls around it, the height of the plant from concrete-loading roadway to tipple on top of the immense storage bins was actually that of a ten-story building. Its equipment was of the most modern type. When placed in operation, sand, the various sizes of gravel, water and cement were fed first into separate batchers which automatically filled to the designated weight, then passed to a mixer-hopper and into the mixer. After a 2½-minute mixing period, the resultant concrete was dumped into buckets and transported to the pouring site. The weights of cement, water, sand, gravel, and the relative amount of water (consistency) of each concrete batch were automatically recorded.

Water was secured from the river, desilted in a settling basin and pumped to storage tanks near the plant. Cement was purchased by the government in bulk, shipped to the plant in box cars, and loaded into bins at the top of the plant by compressed air pumps. Sand and gravel were dug from a deposit on the Arizona side of the river, ten miles upstream from the damsite, loaded by a 5-cubic-yard electric dragline into trains of ten 50-ton dump cars each, and hauled across seven miles of standard gauge line, including an 800-foot pile trestle bridge over the river, to a screening plant in Hemenway Wash near the present boat landing. Here the screens separated the pit run into sand, three sizes of gravel, and 3-inch to 9-inch cobbles, the latter for dam concrete only. Belt conveyors transported the classified materials to stock piles and later they were hauled by train direct to mixing plant or to storage piles.

COFFERDAMS

By November 13, 1932, eighteen months after the first excavation, the two Arizona tunnels were ready to carry the river flow, and simultaneous blasts on that day at inlets and outlets signalized the first diversion. Preceding that event, a pile trestle bridge had been thrown across the river a short distance downstream from tunnel inlets and as soon as a part of the river started

The Lo-Mix concrete plant was located near the river's edge and upstream from the damsite

pouring through the Arizona tunnels, trucks commenced dumping earth and rock on each side of the bridge to form a temporary barrier and force the entire flow out of the old river bed. A dam of similar material was pushed out from the Arizona to the Nevada side upstream from the tunnel outlets and the water entrapped between the two barriers was removed by pumping.

Protected by the temporary banks, fifty trucks, a hundred dump cars pulled by a half dozen engines, tractors, rollers and power shovels toiled without cessation day and night to raise the permanent cofferdams in advance of river floods which might overtop the impermanent dikes. Trainload after trainload was hauled from suitable

The river was detoured through the canyon walls and a temporary dam was built across the channel immediately downstream from the diversion tunnel inlets

The upper cofferdam, a barrier 750 feet thick, 98 feet high and 480 feet wide, composed of earth and rock with its upstream face covered with concrete, turned the river into the diversion tunnels while the dam and powerhouse were built

The lower cofferdam was built upstream from the tunnel outlets to prevent backflow, and the floods of the following spring were easily diverted by the four tunnels

points in Hemenway Wash, five miles from the damsite, dumped into receiving piles, reloaded into trucks, hauled to the cofferdam sites, dumped, moistened, spread by caterpillar tractors and rolled compactly. Rock secured from nearby excavations was placed on the slopes toward the site of the dam. The upstream face of the upper cofferdam was protected by four acres of concrete paving, and the downstream face of the lower cofferdam by a barrier farther downstream of 98,000 cubic yards of rock. A third of a million cubic yards of fill were placed in one month and 18,000 yards in one day. A month's bill for gasoline used by the trucks amounted to $45,000.00 and tires $15,000.00.

Five months after starting work, the upper cofferdam stretched its length of 450 feet across the channel, rising 98 feet above foundation and extending up and downstream 750 feet at the base and 75 feet at the crest. Downstream, the cofferdam of similar length, 66 feet in height, 500 feet in base width and 50 feet at the crest was ready to turn back the eddy flow from diversion tunnels. After their use for diversion had been fulfilled, the lower cofferdam and rock barrier were removed, as otherwise they would obstruct the flow from the powerhouse tailrace.

Designed to safely carry the maximum flood recorded in Black Canyon of 200,000 cubic feet per second, the diversion works were only required in the following years to turn a flow of 70,000. With all four tunnels open for the latter discharge, the water depth at the tunnel outlets was 45 feet. A 200,000 second-foot flood would have raised the water surface eight feet above the tunnel arches.

HIGH SCALING

Before excavations for the dam and powerhouse could be started in the river channel between the two cofferdams, the stripping of loose and projecting

Both canyon walls were stripped of loose rock by the "High Scalers"

Cuts in the canyon walls for all structures were excavated by high scaling methods

rocks from above the sites of all structures and the cutting of niches in the cliffs for intake towers, dam abutments, back walls of powerhouse, and canyon wall outlet works, had to be completed. This work, probably the most hazardous and undoubtedly the most spectacular of all construction proceedings, was of necessity performed by men of iron nerve and dauntless courage. Ropes were secured to steel rods, inserted in drilled holes near the canyon rims, and dangled over the sheer cliffs toward the canyon floor far below. The "Highscalers" climbed down the ropes or were lowered into place by cableway and fastened their safety belts or bosun chairs to the suspended ropes by a special knot. Here, hour after hour amidst the dust from drilling, searing heat from black walls, or raw gravel-laden winds of winter, they drilled for blasting, loaded holes with dynamite, or pried off loose rock, while below lay an unobstructed fall of hundreds of feet. It is to the great credit of the high scaler's skill and his observance of safety precautions that only seven men lost their lives in this dangerous undertaking, although the work was in progress for more than two years and 400 were employed at one time. Rock removed by high scaling methods amounted to 137,000 cubic yards, equivalent to the volume of a slab one foot thick, ten city blocks in length and the height of the Empire State Building in New York City.

DAM

Cuts in the canyon walls for the dam abutments were made on radial lines of the dam axis at the top, where the water thrust against the dam will be carried primarily by arch action, and

Smoke and fumes from blasting filled the canyon with a dense white fog

Preparatory to pouring concrete and placing other materials in the dam and other structures along the canyon, a group of heavy duty 25-ton capacity cableways were strung from rim to rim of the canyon at strategic points. These were five in number and were so located and devised that supplies could be deposited at any point on the dam, spillway open channels, intake towers and most of the powerhouse. All of the cableway towers, at terminals of the 3-inch diameter track cables, were movable except the one stationary head tower farthest downstream on the Nevada side. The movable end structures were built of steel. The largest were 32 feet by 46 feet in plan, as high as a seven-story building, and the span across the canyon nearly a half mile.

The towers traveled on parallel tracks and the end structures moved in unison, being operated by electric motors controlled by the same circuit. The tendency of a tower to overturn was counteracted primarily by million-pound

were warped to the normal canyon wall in a drop of about 400 feet, the dam section at the base being designed to withstand the reservoir pressure by the weight of the structure. Excavations in the river channel were conducted primarily by power shovel and truck, very little blasting being necessary as most of the material in the river bottom was silt, sand, gravel, and small boulders. Bedrock was encountered forty feet below the river bed, except in a narrow middle gorge of eighty-foot width, which extended downward nearly an additional hundred feet. According to geologists, this inner ravine probably was cut at the end of the last ice age. A 2-inch by 6-inch saw-mill timber was found at the edge of the gorge, indicating that in the last 50 years a tremendous flood had scoured out the river bed to the forty-foot depth.

Rock shot down from canyon walls was removed by power shovels and fleets of trucks

Excavations for the dam and powerhouse proceeded downward below the river bed, toward bedrock

concrete blocks, placed above the rear trucks, and by raising the front rail of the front track or by a central truck bearing against an anchored central rail laid with web horizontal. All five cableways were operated from the head towers located on the Nevada side and were controlled by telephonic orders from men in view of the loading or unloading activity. In their few years of use they were required to convey more than eight million tons of supplies and materials.

Another concrete mixing plant was erected near the Nevada canyon rim, 600 feet downstream from the top of the Nevada dam abutment. Equipment in this plant named "Hi-Mix" was similar to that in "Lo-Mix," previously described, but here there were as many as six 4-cubic-yard mixers in operation at one time. A cement blending plant was also built near the Hi-Mix plant. Contracts for supplying cement were in quantities of 350,000 to 1,500,000 barrels each and shipments were received under one contract from as many as five companies. The purpose of the blending plant was to combine cements of different chemical characteristics or to mix the same type from various factories to secure a uniform product.

By June 7, 1933, the exposed bedrock

A narrow gorge, cut perhaps at the end of the Ice Age, was laid bare in the river channel. The downstream face of the upper cofferdam is seen in the background

Preparatory to pouring concrete at locations not readily accessible from Lo-Mix, another mixing plant, termed "Hi-Mix," was erected on the Nevada canyon rim

had been cleaned and the first forms were under construction. On June 6 the first 16-ton (8-cubic-yard) bucket of concrete for the dam was mixed at Lo-Mix, transported by train beneath one of the cableways, picked up, swung part way across the canyon, lowered into one of the forms and dumped at the signal-man's order. Other trains, cableways and the Hi-Mix plant were brought into action, a 20-ton derrick with 138-foot boom was later added for transference of concrete from Lo-Mix, and by August 31 the middle gorge was filled; seven months after starting a million yards had been poured; and all blocks were at crest elevation on March 23, 1935. The average placement in 24 hours was approximately 5,300 cubic yards, and more than 8,000 yards were poured in that time.

Completion of all major features of the dam structure took place before the summer of 1935, thus in less than two years, 1,200 men and modern machinery had placed 3,250,000 cubic yards of concrete, a greater volume than that of the Great Pyramid of Egypt, which, according to Herodotus, required the services of 100,000 men for 20 years in its construction.

As completed, the dam arches upstream on a 500-foot radius, is 727 feet high to walkway on the crest, 660 feet thick at the base, 1,244 feet long and 45 feet thick at the crest, and will raise the water surface 584 feet. Its weight is more than 5,500,000 tons. For comparison the dam is 1½ times higher than the Los Angeles City Hall and from base to top of towers only 32 feet less than the Woolworth Building in New York City. The base thickness is more than two normal city blocks and its

A group of cableways with traveling end towers were installed to transport concrete and other materials from canyon rim to pouring site

Bedrock was cleaned, forms were built, and on June 6, 1933, the first concrete was placed in the dam. The view is toward the downstream cofferdam which may be seen in the background

A bucket of concrete was picked up from the train by cableway and carried out over the canyon

Each bucket contained 8 cubic yards of concrete weighing 16 tons

crest more than four blocks in length. The dams of nearest height are Shasta, 602 feet, in Northern California, and Grand Coulee, 550 feet, in the State of Washington.

Placing of this huge mass of concrete in a relatively confined space introduced problems of contraction from the cooling of the concrete, with resultant forming of cracks, that required much

The dam was built as a group of vertical columns closely keyed together

Nine months after the first bucket of concrete was poured, the dam had reached a ***height of 300 feet***

A crane was installed to supplement the cableway operations. This view is from the top of the Nevada abutment and toward the upstream face

Work proceeded both night and day

thought and experimentation before a satisfactory solution was reached. The temperature of the concrete at the completion of setting would be as high as 138° F., and if the dam were allowed to cool naturally, more than 150 years would elapse before the concrete resumed the temperature of the surrounding medium, during which time cracks would develop throughout the structure due to unequal contraction. The plan finally agreed upon was to build the dam in a series of 230 vertical columns, 25 feet to 60 feet in plan, key all sides of columns to adjacent ones, place 1-inch tubing at approximately 5-foot intervals both vertically and horizontally in the concrete while it was being poured, and pass water at a temperature as low as 38° F. through the tubing until the temperature of the surrounding concrete was lowered to the desired degree, which varied from 43° near the base to 72° at the crest. The consequent contraction due to cooling opened up spaces between the columns. These were filled with a water-cement mixture of grout through ½-inch diameter pipe that had been placed as concreting progressed. Cooling of water was accomplished by a 3,000-gallon per minute evaporation tower and an ammonia refrigeration plant, which if turned to ice production could have produced 1,000 tons in 24 hours. These structures were located on or near the lower cofferdam. Fourteen inch pipe carried the cooling water to a slot of eight-foot width in the center of the dam, where connections were made to the smaller cooling headers and tubing. Pipe placed in the dam amounted to 580 miles for cooling and 200 miles for grouting. Water flowed through the cooling loops at the rate of four gallons per minute. The structure was cooled in a period of 20 months and grouting finished shortly thereafter.

Installation of technical instruments in the dam included 410 electrical resistance thermometers, 385 strain meters, 91 contraction joint meters, a system of uplift pressure pipes and gauges, and a group of tiltmeters, these latter to measure the movement of the dam in an up and downstream direction from temperature and water pressure changes.

Shafts and galleries for grouting, drainage and inspection purposes, two miles of them, pierce the structure at various levels in circumferential and radial lines. One of these starts near one abutment, follows it downward within 5 to 30 feet of rock, crosses the dam near bedrock and ascends the opposite abutment. Elevators carry passengers from the two middle towers on the dam crest to the different gallery levels, stopping finally 528 feet down, where there are two, beautifully tiled, which lead downstream through the dam a distance of a block and a half to the central portion of the powerhouse.

Both abutments and the dam foundation were thoroughly drilled to depths as great as 150 feet and grout forced into the rock under pressures as great as 1,000 pounds per square inch. Drainage holes of 3-inch diameter were then drilled, as much as 100 feet deep, from the abutment gallery into the rock downstream from the aforementioned grout curtain to intercept any water passing the dam through the canyon walls.

SPILLWAYS

Flow of water from the reservoir will normally be regulated by the amount required for the turbines of the power plant, or be by-passed around the plant through the needle valves in outlet works to satisfy downstream demands or to lower the reservoir surface. Under exceptional conditions when the reservoir is nearly full as a result of unusually large floods, water will pour over weirs into spillway channels, plunge 500 feet downward through inclined shafts to outer diversion tunnels and emerge in the river channel below the dam and powerhouse.

Twice the maximum flow of water

Spillways were constructed at locations upstream from the dam and near the elevation of its crest to take the overflow from the reservoir

Aerial view of Ho

, looking upstream

BUREAU OF RECLAMATION PHOTO BY JOHN SANTA

recorded passing Black Canyon, or 400,-000 cubic feet per second could be passed by the spillways. More than 11,000,000 horsepower in terms of falling water would be released at this capacity and the velocity of flow as the water rushed into the diversion tunnels would be at the rate of 175 feet per second or 120 miles per hour.

The length of a spillway is approximately 650 feet in open section and the concrete lined channel is 150 feet wide at the top and 125 feet average width. Comparatively speaking, the channel length is greater than two blocks, width a half block, and height of channel wall as great as a 10-story building. The largest battleship could be floated in the structure if the inclined shaft were dammed at the portal. The Arizona shaft is 88 feet high and 98 feet wide at the portal and narrows gradually to a 50-foot diameter. A concrete plug, 400 feet long, was placed in the diversion tunnel immediately upstream from the intersection with the inclined shaft. Four sets of 6½ by 7-foot slide gates were installed, at first, in the Nevada plug, for the purposes of regulated diversion, while the reservoir penstock system and outlet works were made ready to carry water from the reservoir.

The outlet for each spillway was through a 50-foot diameter inclined shaft which connected with an outer diversion tunnel

Drum gates were erected on the spillway crests to regulate the reservoir overflow

Four steel drum gates are erected on each spillway crest. These are structural, buoyant vessels which rest in recesses in crests between the piers, and are raised and lowered, respectively, by filling the recesses with water from the rising reservoir, or emptying them into the spillway inclined shaft. Each gate has a flat bottom and two curved sides resembling in section when in raised position, the quadrant of a circle of 17 feet radius having a 3-foot extension projecting upward from the reservoir face. The length of a gate is 100 feet, its height when raised is 16 feet above the permanent crest, and weight a half million pounds. In lowered position the reservoir side of the gate provides a curved surface to complete the outline of the weir crest. The permanent crest of the spillway is about 27 feet below the dam crest and the top of the gates

A highway bridge spans the Arizona spillway. To gain an impression of the structure's dimensions, note the figure of the man in the spillway channel

Attention is directed in the picture to the spillways on each side of the canyon, and the four intake towers upstream from the dam

when raised, approximately 11 feet below.

The spillway channels were blasted from solid rock, using the drill jumbos from the tunnels to a certain extent, and removing the rock by power shovels and trucks. The inclined shafts were excavated by first driving top headings from the diversion tunnels upward, enlarging these into slots entirely across the shaft section, and breaking down the remaining side sections into the slots working in horizonal benches from the top downward. Muck was removed from the diversion tunnels by power shovels and trucks.

Lining the channels with concrete from Hi-Mix was performed with the aid of wooden forms, and transportation by truck, cableway, and draglines equipped with 100-foot booms. Concrete was taken to the wooden forms in the inclined tunnels via belt conveyors and chutes from agitators (resembling transit mixers) brought down from the

The bases of intake towers are in niches cut in the canyon walls, 260 feet above the old river bed

The towers control the flow of water from the reservoir to the power plant and outlet works

shaft portals by rail.

INTAKE TOWERS

Water for the production of electrical energy, supplying of downstream requirements and normal regulation of the reservoir, flows through the gates of intake towers into penstock headers and penstocks to the turbines of the power plant, or from the penstock headers into outlet headers and outlet conduits to the needle valves in outlet works.

The four towers that rise in the reservoir immediately upstream from the dam are notable examples of the possibilities presented for building beautifully as well as for strength and utility. These graceful concrete spires resemble huge fluted columns and have all the appearance of monuments rather than serviceable structures.

Actually each tower is a hollow concrete cylinder of 29 feet 8 inches internal diameter and 75 feet average outside diameter from which 12 fins project radially, the openings between the fins being spanned by steel trashracks to

Cylindrical gates, 32 feet in diameter, have been installed in the bases of all intake towers and at locations 150 feet higher. These are raised and lowered vertically by hoists in the tops of the towers

prevent debris from entering gate openings. A hoist house of more than four stories height sits atop the tower and contains electrically operated hoists for raising and lowering the cylindrical gates that are installed, one at the tower base and another 150 feet higher. The gates are 32 feet in diameter, eleven feet high, and weigh approximately a half million pounds each. Any gate can be isolated for repairs by steel bulkheads which may be lowered down grooves in the trashrack fins.

Two towers are in each side of the canyon, the downstream ones being 135 feet from the dam face and the other two, 185 feet farther upstream. Bridges join the upstream and downstream towers with each other and with the dam. The overall height of a tower is 395 feet above its rock foundations, equal to that of a 39-story building. Foundations for the structures are on shelves cut in the canyon walls 250 feet above the old river bed thus providing a basin for silt storage and procuring a clear water flow to the powerplant turbines.

Concrete was secured from the Hi-Mix in 4-cubic-yard buckets and transported by rail, or rail and cableway, to the canyon rims above the towers. Huge derricks of 15-ton capacity and 180-foot booms picked up the buckets and lowered them to central hoppers above the completed sections of the towers, from which the concrete flowed through radial chutes to fins, beams and barrels. Heavy reinforcement was used in construction, approximately 150 pounds of steel being placed for each yard of concrete.

PENSTOCK AND OUTLET SYSTEM

The arrangement of tunnels and included steel pipes which carries water from intake towers through canyon walls to powerhouse or outlet works is termed the "penstock and outlet system." In reality, there are four systems, each of which starts from the base of an intake tower, supplies water for four or five turbines, and has an outlet works opening into the river channel at the downstream end of the pipeline. The main supply lines are called "headers," the connections to turbines, "penstocks," and the conduits to outlet works, "outlets."

Starting from bases of upstream towers, inclined shafts of 41-foot diameter rock section and lined with a 2-foot thickness of concrete drop vertical distances of about 200 feet to join the inner diversion tunnels. Plugs, 300 feet in length, completely fill the diversion tunnels immediately upstream from these intersections. Opposite the power-

Header tunnels 41 feet in diameter were driven from downstream intake towers to the canyon wall outlet works

Penstock tunnels run from header and inner diversion tunnels to back wall of powerhouse

Outlet works for the upstream intake towers were located in plugs in the inner diversion tunnels, near their lower ends. Plug excavation is pictured

The steel pipe line header that starts at the base of one of the upstream intake towers ends at the manifold section shown here in the inner diversion tunnel plug

The steel pipe line header that starts at the base of one of the downstream intake towers ends at the group of pipes shown here in the canyon wall valve house

house wings, four tunnels of 21-foot excavated diameter and 18-inch lining of concrete take off from each of the diversion tunnels and extend to back-walls of the powerhouse. Thirty-foot diameter plate steel pipe run from the tower bases down the inclines through the diversion tunnels, past the power-house connecting tunnels, are then reduced to 25-foot diameter sections and end at the needle valves in plug outlet works. A thirteen-foot diameter steel penstock rests in each of the 18-foot tunnels to the powerhouse, connecting with a butterfly valve and one of the turbines.

A plug outlet works consists of a concrete plug 142 feet in length, 85 feet high and 105 feet wide in maximum dimensions situated in each of the two inner diversion tunnels about 600 feet upstream from the tunnel portals, and equipped with six 72-inch needle valves and emergency slide gates. The outlet portals of the two diversion tunnels contain 35-foot by 52-foot Stoney gates of six-foot maximum thickness and weighing 230,000 pounds, which may be lowered to cut off back flow from the river channel while work is done at the plugs.

Beginning at the bases of the two downstream intake towers, tunnels of 41-foot diameter in rock and lined with concrete to 37-foot diameter, drop through a vertical curve of 7-foot radius and extend horizontally downstream through the cliffs. Opposite the power-house, four tunnels of 21-foot excavated section and 18-inch concrete lining leave each of the 37-foot header tunnels and descend on an incline a vertical height of 183 feet to the powerhouse. At the end of each of the header tunnels, six 13-foot horseshoe-shaped bores lead outward through the cliffs to the canyon wall outlet works. In similar manner to the installations previously described, 30-foot diameter pipe extend from the bases of downstream towers past the penstock tunnels to the power-house, and are then reduced to 25-foot diameter sections. Four 13-foot penstocks lead from each of the headers to powerhouse turbines, the line at the downstream end of the Arizona system splitting into two nine-foot diameter pipes to furnish power for two smaller turbines.

The six 84-inch needle valves and emergency slide gates of each canyon wall outlet works connect with the 25-foot pipe section through 8½-foot diameter conduits and an immense pipe manifold. The "cliff dweller" structures of these outlet works are placed in structures cut in the canyon cliffs 160 feet above the old river bed and 200 feet downstream from the powerhouse. Their

Needle valves of 84-inch diameter located in the valve houses supplement the flow of water downstream from the powerhouse turbines

dimensions are equivalent to those of a building 40 feet wide, two-thirds of a block long, and six stories high. A particular feature of note in their construction is the design of the roof structure. The ceiling of each valve house is a reinforced concrete slab twelve inches thick, on whose upper face rest 132 thirteen-inch in height and twelve-inch diameter low-strength concrete cylinders. These in turn support a grillage of railroad rails compactly placed side by side and encased in eight inches of concrete. On top of the rail-concrete slab are eleven inches of compacted gravel covered with one inch of bituminous sand mastic. Thus a large rock striking the roof will at most only crush one or more of the lower-strength columns, which can be replaced with little difficulty.

Excavation and lining of the inclines to intake towers were accomplished by much the same methods as were used for the inclined shafts from spillways, and in the 37-foot horizontal sections of tunnel in like manner to the diversion tunnels. The smaller penstock tunnels were driven by usual tunneling methods employing mucking machines and trucks where permissible. Concrete was placed in the linings with concrete pumps behind movable wooden forms.

Header lines rest on piers and thrust blocks for most of their lengths, allowing continuous inspection, but are anchored by filling the space between pipe and tunnel lining in the inclines to towers, upstream and downstream from penstock openings, at the plugs outlets of inner diversion tunnnels, and the outlet manifolds of the upper systems. The penstocks leading to powerplant turbines are anchored at the powerhouse walls and near the intersections with header pipes.

Four construction adits were also driven into canyon walls from locations immediately downstream from the powerhouse and directly beneath the

Fabrication of penstock pipe was conducted near the canyon, as the sections were too large to be transported on existing railroads

A 24-foot length of 30-foot diameter pipe weighed as much as two standard gauge locomotives

The larger pipe sections were hauled from fabrication plant to canyon rim on a 200-ton trailer

150-ton government cableway. These are 41 by 56 feet in section, and extend to one or the other of the header tunnels. Their purpose was to provide means of access to the tunnels for the large pipe sections.

FABRICATION AND INSTALLATION OF PLATE STEEL PIPE

Most of the pipe sections for the penstock and outlet system were too large and too heavy to be transported by existing railroads, and therefore the Babcock and Wilcox Company, contractor for furnishing and installing the pipelines, erected a plant, of three blocks length and the height of a six-story building, a mile and a half west of the Nevada dam abutment.

Operation of the plant can best be described by following the fabrication of a 30-foot section. Six plates of high tensile steel, each 11 feet wide, 31.5 feet long, and 2¾ inches thick were brought into the plant by rail, unloaded by bridge crane, marked to designated shapes, and welding grooves cut along the edges by a specially designed 50-foot planer. The ends of the plates were bent to the prescribed radius by a 5,000-ton hydraulic press, and the plates then rolled to required form in a vertical press 12½ feet in height. Three of the rolled plates were welded together by automatic electric-arc machines, forming a section 11 feet long and 30 feet in diameter, and then welded to the segment fashioned from the other three plates. Stiffener rings and butt straps were added, and the welded joints subjected to the scrutiny of a 300,000-volt X-Ray, which provided a radiograph of the fusion weld and revealed any defects therein. Any imperfections were chipped out, refilled with new metal and X-rayed again.

The required supporting brackets were added, the section was picked up by one or more of the 75-ton bridge cranes and conveyed to a stress relieving oven, placed therein and the temperature raised to 1150° F. After soaking at this heat for approximately three hours, the section was cooled gradually in about the same time. This procedure relieved the stresses set up in the pipe by rolling and welding. The final act of fabrication was the machining of the pipe ends to exact measurements by a huge facing lathe, equipped with a 35-

Transportation of pipe from canyon rim to portals of construction adits on the canyon walls was accomplished by a cableway capable of handling 200-ton loads

The pipe section arrives at the portal of a construction adit through which it will be moved to position in a header tunnel

foot arm. The interior of the pipe was then shot blasted and coated with coal tar, the exterior was cleaned by air and wire brush and painted. When completed the 30-foot diameter section, 24 feet in length weighed more than 150 tons.

Transportation and erection of the heavy unwieldy pipe sections presented problems as difficult as those encountered in fabrication. Imagine a steel cylinder weighing as much as two locomotives and of sufficient size to encompass a dwelling of three rooms on one floor and an equal number upstairs. Now visualize the moving of this object from plant to canyon rim, lowering it to a bench on the cliff 600 feet below, taking it into the rock wall several hundred feet to a tunnel intersection and moving it through the tunnel as much as a third of a mile. Then it had to be placed in an exact position and joints completed of sufficient strength to withstand the pressure of a 600-foot head of water. This procedure was repeated 222 times for the 30-foot pipe sections and more than 400 times for those of lesser diameter.

All pipe of 13-foot diameter or less were transported by railroad from the plant to the Nevada rim, lowered by the 150-ton Government cableway to portals of construction adits and hauled to position by cars and hoists. The larger sections were loaded on a 200-ton trailer, pulled by two 60 H.P. tractors to the Government cableway, lowered onto specially designed railroad carriages at adit portals, taken sidewise into the header tunnels and pushed endwise into position.

The permanent cableway owned by the government deserves more than passing note. It was by far the largest of its kind ever built, having a nominal rating of 150 tons and being required at times to carry loads of more than 200 tons including track carriage and fall blocks. Six 3½-inch cables at 18-inch centers form the track. Their anchorage was secured by driving tunnels 50 to 80 feet into the cliffs, excavating eighteen foot bulbs at the ends of the tunnels, connecting the cables by plates and eye bars to 13 by 13 foot steel grids in the bulbs and pouring bulbs and tunnels full of concrete. The span is 1,200 feet, maximum lift over 600 feet, hoisting speed 40 feet per minute with heavy loads and 140 feet per minute with lighter loads, and conveying speed is 240 feet per minute. Operation is by remote control from the station on the Nevada rim near the upper loading platform.

Having arrived in the tunnels, the pipes were joined together, generally, by heating the butt strap on one end of a pipe by a gas ring, and pushing the end of the adjacent pipe within the strap. The butt strap and included pipe were then fastened together by rivets or pressure pins (these latter as much as three inches in diameter and seven inches long) and the outer end of the butt strap caulked tight.

The interior of a 30-foot diameter header at the intersection of one of the 13-foot diameter penstocks

A thirty-foot diameter header was erected in each of the inner diversion tunnels. The picture was taken, looking downstream through the diversion tunnel, from the raise that leads to an upstream intake tower

Closure of the last joint left open between any two anchors was made after the pipe had been pre-stressed by placing jacks at the closing joint and contracting the pipeline to the same extent as would have been obtained by a drop in temperature to 54°F. Water from the reservoir has approximately this temperature, consequently the temperature stresses in the pipe are reduced to a minimum by this pre-stressing at the time of erection.

The weight of the fabricated pipe exceeded 88,000,000 pounds or more than 900 car loads. Approximately 76 miles of electric welding were required, using 1,000 tons of welding rod. Film for the X-ray photographs if placed end to end would extend 26 miles.

POWERHOUSE

Immediately downstream from the dam lies the huge concrete and steel

The powerhouse is a U-shaped structure at the downstream face of the dam

structure of the powerhouse, a U-shaped building whose two wings nestle close to the cliffs, and the central section connecting them lies on the downstream face of the dam. The length around the U next to the cliffs and dam is nearly that of six ordinary residence blocks and the average width of each wing or central section approximately a half block. Its height from lowest concrete to top of highest parapet is that of a 20-story building, and the parapet rises the height of 12 stories above low water surface in the tailrace. The roof covers an area of four acres, is 4½ feet thick (to resist rock falling from the cliff above) and is composed of seven laminations, two of these being reinforced concrete, another asphalt paving, and others of sand and gravel. Support for the roof is provided primarily by 11,600,000 pounds of structural steel, including 5 miles of 33-inch I-beams and 88 trusses 67 to 73 feet in length. Beneath the roof are 10 acres of floors.

When finally completed, the powerplant will contain fifteen 115,000 horsepower units, and two of 55,000 a total installed capacity of 1,835,000 horsepower. Included in the plant machinery are 300-ton bridge cranes, 14-foot diameter butterfly valves, turbine scroll cases 40 feet across, 82,500 KVA generators, and 55,000 KVA power transformers for raising the generator voltage from 16,500 to transmission voltage as high as 287,500 volts. The building contains nearly a quarter of a million cubic yards of concrete, 22,000,000 pounds of reinforcing steel and eleven miles of pipe and conduit. Operating at rated capacity, the plant would be capable of producing sufficient electrical energy to supply complete domestic light and power for all the eight and a half million inhabitants of the Colorado River Basin,

Water will enter the spiral scroll case of the 115,000 horsepower turbine, pass between the vanes in the inner rim, and strike the horizontally rotating turbine runner

Electricity is generated by rotation of the rotor field in the stator windings. Rotor of an 82,500 KVA generator is shown on the right, and stator on the left

View in the Nevada wing of powerhouse showing initial installation of four generator units. Small station generator in foreground

or, calculated in a different manner, it would be enough to furnish each and every home in the United States with light from a 40-watt bulb.

Access to the powerhouse is gained by two elevators, which descend from the dam crest a distance equal to the height of a 44-story building, and then by passageways through the dam a block in length to the central sections. Another route for truck haulage is by a mile and a half of road and a 1900-foot tunnel which connect the Boulder City highway with the downstream end of the Nevada powerhouse wing.

Operation of the plant may be more readily understood by following in a general way the transformation of falling water to electrical energy and the route of the electrical current from one of the larger generators to a transmission line. Water from the reservoir flows through intake tower gates into a 30-foot diameter penstock header and down a 13-foot diameter penstock to the powerhouse. Passing through a 14-foot butterfly valve at the rate of 2000 to 3000 cubic feet per second it enters the spiral scroll case of the turbine. Twenty-four speed vanes and wicket gates in the inner rim of the scroll case automatically regulate the amount of water striking the turbine runner, where the water's energy is transformed into mechanical energy by rotating the horizontal runner and its vertical shaft. With its velocity greatly reduced, the water passes outward to the powerhouse tailrace through the center of the turbine and connected draft tubes. The mechanical energy is transmitted by vertical shaft to the generator rotor, which revolving in an electromagnetic field generates electricity. The electromagnetic field is created by two generators, termed exciters, mounted above the main generator and rigidly connected on the main shaft.

Electrical energy leaves the 3-phase, 60-cycle, 82,500 KVA generator terminal at 16,500 volts, passes through oil-circuit breakers to 23,000 volt 4,000 ampere bus structures and then flows to three 55,000 KVA single phase transformers outside of the powerhouse where it is raised to 287,500 volts for transmission

The initial stage of construction was compnleted on November 13, 1932, when the river was first diverted around the damsite through the canyon walls

purposes. From the transformer it is carried up and out of the canyon on an overhead high voltage circuit to a switchyard located on the Nevada side of the canyon and approximately 1,500 feet from the canyon rim, where it passes through oil circuit breakers and disconnecting switches to the transmission circuits leading, in the example just described, to Los Angeles. Remote control from the powerplant, of the oil circuit breakers located at the switchyard, is provided by electrical circuits running from the central section upward through the cliff in an inclined shaft and thence continuing to the switchyard in a 6-foot by 8-foot concrete conduit.

Governors controlled by the speed of the turbine runners, and electrically and then mechanically connected to turbine wicket gates, regulate the flow of water through the turbine. In case of an outage in the electrical circuit requiring an emergency shutdown, water will be automatically by-passed around the turbine through a pressure-regulator valve and a relief valve. Each generator may be operated from a control station near it or by a main control board on the top floor of the central section of the powerhouse.

SUMMARY OF CONSTRUCTION

Almost as astonishing as the magnitude of construction was the speed at which the work was accomplished. First appropriations were granted July 3, 1930, the contract was awarded Six Companies, Inc., on March 11, 1931, and initial excavations for diversion tunnels took place May 16, 1931. The river was detoured from its channel November 13, 1932, cofferdams completed April 1, 1933, and the first concrete poured the following June 6. Storage of water was started February 1, 1935, when the 2,-000,000 pound bulkhead gate at the inlet of the Arizona outer diversion tunnel was lowered, turning the river flow through the outer tunnel on the Nevada side, under control of slide gates in the plug located immediately upstream from the connection with the spillway inclined shaft. All blocks in the dam were raised to crest elevation on March 23, 1935, and the construction completed by Six Companies, Inc., was accepted by

Storage of water in the reservoir was started on February 1, 1935, when the 2,000,000-pound gate was lowered at the inlet of the Arizona outer diversion tunnel

the Bureau of Reclamation on March 1, 1936, more than two years in advance of the scheduled date of completion. Electrical energy was produced by the powerplant in the Fall of 1936.

The dam was dedicated on September 30, 1935, by President Franklin D. Roosevelt in a ceremony which was attended by Secretary of Interior, Harold L. Ickes, several Senators and Governors from the Colorado River Basin States, high officials of the Bureau of Reclamation, and many prominent persons whose names are linked with the Colorado River conflict and the development of the Southwest.

Many construction records were established by the aid of equipment of improved design and unusual size. As much as 256 lineal feet of 71 by 56-foot tunnel heading were driven in one day, breaking 17,000 cubic yards of rock. Day after day the sand and gravel screening plant classified 800 tons of concrete aggregates per hour, and at times 1,000 tons passed through the plant in a 60 minute period. More than 10,000 cubic yards of concrete were produced by the two mixing plants in a day, requiring 35 carloads of cement, 318 cars of aggregates and 270,000 gallons of water. One of the 25-ton cableways transported 370 8-cubic-yard buckets of concrete to place in the dam in eight hours.

Excavation, mostly in rock, amounted to eight million cubic yards, which if placed in a masonry wall similar to those on the project, would extend across the United States between the Atlantic and Pacific oceans. Included in this work was the driving of approximately 110 tunnels and shafts whose aggregate length was more than seven miles. Concrete placed in construction totalled nearly four and a half million cubic yards, sufficient to build a 20-foot pavement from Florida to California, or a 5-foot sidewalk from the North to the South pole.

Supplies for building the huge structure included 5,000,000 barrels of cement, 165,000 cars of sand, gravel and cobbles, 35,000 tons of structural and reinforcing steel, 900 cars of hydraulic machinery and over 1,000 miles of steel pipe. All states in the Union sent supplies to form a part of the completed works, the requirements ranging from tin cups and friction tape to steel gates weighing three million pounds apiece

The dam was dedicated by President Franklin D. Roosevelt on September 30, 1935

and generators of 2,300,000 pounds each. Forty railroad cars were used to transport each of these gates and specially built low-center cars of 75-ton capacity were required for generator parts and transformers. If the total materials and construction equipment used on the project were placed in one train the engine would be arriving in Boulder City as the caboose left Kansas City, Missouri!

RESULTS

The vision held for so many years of adequate protection for downstream lands from drought, floods, and silt has materialized, and the hum of the generators in the power plant are inaugurating a new era of prosperity and development in the Southwest. Fears of Imperial Valley dwellers now may be allayed, for several years' supply of irrigation water is in the reservoir, and floods above the dam may be shut off entirely if found necessary. The silt content of the river water at downstream points has been lessened in large degree and will continue to be reduced as the river channel becomes stabilized.

Complete control of the Colorado was secured in June, 1936, when all the reservoir discharge passed through the needle valves of the Arizona canyon wall outlet works

All transmission lines converge at the switchyard, which is located west of the Nevada canyon rim

Rows of tall towers carrying lines of copper and aluminum now cross the desert, bringing millions of watts of power to Southern California, Western Arizona and Southern Nevada. The transmission lines of the Los Angeles Bureau of Power and Light are the largest ever erected. Two rows of towers 109 feet high and spaced 800 to 1,000 feet apart carry the conductors a distance of 230 miles, from the powerplant to Cajon Pass, and single towers 144 feet in height—nearly that of a 12-story building—carry the two circuits the remaining 40 miles to Los Angeles. The Heddernheim type of conductor was adopted for use in these circuits after extensive research. This is a hollow-core copper tube of 1.4 inches outer diameter made up of interlocking spiral segments. Work was started on the line in June 1933, following a loan of $22,800,000 from the Reconstruction Finance Corporation to the Bureau of Power and Light, and completed three years later. Among the more important items of construction are 27,000 tons of structural steel, 1,600 miles of conductor, 1,000 miles of counterpoise and 254,000 porcelain insulators of 10-inch diameter.

Lines of tall towers are marching across the desert, on which conductors will be strung to carry power to Southern California Cities

Other projects which were impractical to construct until Boulder Dam was built, include the $220,000,000 Metropolitan Aqueduct to take water from the river at Parker Dam and carry it 400 miles to Los Angeles and nearby municipalities, and the $38,500,000 All-

Passenger boats make scheduled trips at frequent intervals from Regatta Bay to the upstream face of the dam

American canal to supply the irrigation needs of the Imperial and Coachella Valleys.

One of the most striking transformations brought about by the storage of water in the reservoir, officially named Lake Mead, is the change in color and consistency of the Colorado River. Where once it flowed through Black Canyon a brown turbid stream, carrying an average silt load of 300 tons a minute—"too thin to plow and too thick to drink"—the lake now is a dark emerald green in the canyons and a deep blue in more open country, while the river below the dam has regained the sparkling clearness it possessed when it left the mountains.

An unusually scenic and interesting country has been made accessible by the filling of the canyons where the river long held sway. The fiord-like vistas of Boulder, Black, and Iceberg Canyons and the lower reaches of Grand Canyon, the many islands inhabited by desert dwellers secluded there by the rising waters, the deep colors and mirror-like reflections of unusual rock formations, and the ferns, springs, falls and prehistoric dwellings in remote side canyons, are among the sights viewed on exploratory boat trips. Shorter excursions are made at frequent intervals and there are regular schedules from the boat landing to the upstream face of the dam and nearby places.

The lake is being stocked with millions of bass, crappie and bream. Trout also have been placed in the river downstream from the dam, as the temperature of the water will be cool both summer and winter, being drawn from a reservoir depth of 150 to 300 feet.

The fiord-like vista of Boulder Canyon is among the views seen on exploratory trips up the lake

A mid-winter scene. The richly colored Temple is mirrored in the dark blue waters of the lake

PERSONNEL

Practically all construction was completed during the administrations of President Herbert Hoover and President Franklin D. Roosevelt. The head of the Department of the Interior during each of these terms was, respectively, Secretary Ray Lyman Wilbur, and Secretary Harold L. Ickes.

Principal officials of the Bureau of Reclamation and contractors who were in charge of project activities during most of the construction are listed below:

Bureau of Reclamation

Commissioner Elwood Mead, Washington, D. C.

Denver, Colorado, Office

Chief Engineer, R. F. Walter.
Asst. Chief Engineer, S. O. Harper.
Chief Designing Engr., J. L. Savage.

Boulder City, Nevada, Office

Construction Engr., Walker R. Young.
Office Engineer, John C. Page.
Field Engineer, Ralph Lowry.
Chief Clerk, E. R. Mills.
City Manager, Sims Ely.

Six Companies, Inc.

Presidents: W. H. Wattis.
W. A. Bechtel.
E. O. Wattis.
H. W. Morrison.
Director of Constr., Charles A. Shea.

Boulder City, Nevada, Office

Gen. Superintendent, Frank T. Crowe.
Asst. Superintendent, B. F. Williams.
Chief Engineer, A. H. Ayers.
Administration Manager, J. F. Reis.

The Babcock & Wilcox Company

President, A. G. Pratt, New York City.

Barberton, Ohio, Office

Vice-President, Isaac Harter.
General Superintendent, J. E. Trainer.

Boulder City, Nevada, Office

Project Superintendent, B. T. Kehoe.

ACKNOWLEDGMENT

Grateful acknowledgment is hereby made to the Bureau of Reclamation, Six Companies, Inc., the Babcock & Wilcox Company, and The Los Angeles Bureau of Power and Light for photographs supplied and assistance rendered in the preparation of this booklet.

An aerial view looking northwest to Hoover Dam and Lake Mead. In the background is Boulder Basin, one of the larger basins in the reservoir behind Hoover Dam.

DO YOU KNOW THAT?

Construction of the Project involved:

The driving of 110 tunnels and shafts, whose aggregate length is more than seven miles.

Placing of sufficient concrete to build a walk five feet wide from the North to South pole.

Erection of 2,000,000-pound gates, 14-foot Butterfly valves, 7-foot needle valves.

Furnishing of a sufficient quantity of materials to form a train from the dam to Kansas City, Missouri.

The Spillways:

Are each large enough to float the greatest battleship, have a channel height equal to a 10-story building.

When operating at capacity would pass twice the recorded river flow, release 11,000,000 horsepower as falling water, discharge at a velocity of two miles a minute.

The Dam:

Was the highest in the world.

Has a larger volume than the Great Pyramid.

Contains two miles of shafts and galleries.

Was cooled to temperatures considerably below atmospheric while being built.

The construction plant contained:

Fifty-ton trucks, 150-passenger lorries, a 39-mile railway system, a mess hall that served 6,500,000 meals, cableways of 25-ton capacity whose end towers traveled along the canyon rims, a cableway capable of transporting 200-ton loads.

Each Intake Tower:

Is as high as a 34-story building.

Contains two cylindrical gates, 32 feet in diameter.

Steel pipe in lines to powerhouse and outlet works:

Weighs 88,000,000 pounds, required 76 miles of welding and 1,000 tons of welding rod in construction.

Was X-rayed at welded joints, requiring 26 miles of film.

Has a maximum diameter of 30 feet, thickness of 2¾ inches, and a 24-foot length weighed as much as two steam locomotives.

Lake Mead, when filled to capacity:

Is 500 feet deep and 110 miles long.

Contains enough water to cover Connecticut 10 feet deep or to supply 5,000 gallons to every person in the world.

INDEX

The publisher extends thanks to the Visitors' Bureau of Boulder City, Nevada, for its help in making this account available for publication.